26th Annual Saudi-Japan Symposium on Technology in Petroleum Refining and Petrochemicals 2016

AA005995

Dhahran, Saudi Arabia
7 - 8 November 2016

ISBN: 978-1-5108-3306-7

Printed from e-media with permission by:

Curran Associates, Inc.
57 Morehouse Lane
Red Hook, NY 12571

CURRAN ASSOCIATES INC. proceedings .com

Some format issues inherent in the e-media version may also appear in this print version.

Copyright© (2016) by King Fahd University of Petroleum & Minerals
All rights reserved.

Printed by Curran Associates, Inc. (2017)

For permission requests, please contact King Fahd University of Petroleum & Minerals
at the address below.

King Fahd University of Petroleum & Minerals
Dhahran 31261
Kingdom of Saudi Arabia

Phone: +966 (13) 860 0000

info@kfupm.edu.sa

Additional copies of this publication are available from:

Curran Associates, Inc.
57 Morehouse Lane
Red Hook, NY 12571 USA
Phone: 845-758-0400
Fax: 845-758-2633
Email: curran@proceedings.com
Web: www.proceedings.com

Contents

ارامكو السعودية
Saudi Aramco

الندوة السنوية السعودية اليابانية السادسة والعشرون

26th Annual Saudi-Japan Symposium

التقنية في تكرير البترول والبتروكيماويات

Technology in Petroleum Refining & Petrochemicals

PROCEEDINGS: 26

٧-٨ صفر ١٤٣٨هـ

November 7-8, 2016

KFUPM Research Institute

King Fahd University of Petroleum & Minerals

Dhahran, Saudi Arabia

Preface

This is the Proceedings/Abstracts of the 26th Annual Symposium "*Technology in Petroleum Refining & Petrochemicals*" held on November 7-8, 2016 at the Research Institute of King Fahd University of Petroleum & Minerals (KFUPM), Dhahran, Saudi Arabia. This event, jointly organized by KFUPM, Saudi Aramco, the Japan Petroleum Institute (JPI), and Japan Cooperation Center, Petroleum (JCCP), brings together researchers from industry and academia to discuss technologies related to refining and petrochemicals.

At this year's symposium, there are 23 presentations in six sessions featuring papers on modeling, catalytic reactions, hydroprocessing, aromatics, petcoke, catalytic olefins, cracking, dehydrogenation, gas storage and carbon steels-corrosion. Among the distinguished speakers from Japan, we have experts from the universities of Tohoku, Mie, and Hokkaido as well as from Non-Destructive Inspection Co., JX Nippon Oil & Energy and Chiyoda Corp. We also have highly-cited Saudi Aramco chair professor from the University of Delaware. For local participation, we have distinguished speakers from Saudi Aramco, SABIC and various centers at KFUPM.

We extend our thanks to Saudi Aramco, KFUPM, JPI, JCCP, distinguished speakers, and attendees for their active participation and look forward to your participation in our next symposium.

Gap in pagination due to formatting issues.

Pages 1-2

Paper No. 1

Software Tools for Molecular-level Kinetic Modeling of Refinery and Petrochemical Reactors

Michael T. Klein
Center of Research Excellence in Petroleum Refining & Petrochemicals, King Fahd University of Petroleum & Minerals and Energy Institute, University of Delaware, mtk@udel.edu

Extended Abstract

A chemical kinetic model is a set of mathematical equations that describes the dependence of the rate of chemical reactions on process variables such as temperature, pressure and reaction composition. At the molecular level, there will be one equation for each species in the reaction mixture. For the most important chemical reactors used in the chemical and refining industries, the equations will generally be ordinary differential equations (ODE's), and numerical integration of these ODE's provides the model predictions of the composition of the species as a function of contact time in the reactor.

The considerable interest in Molecule-Based kinetic Models (MBM) for the complex chemistries in the petroleum refining industry is motivated by the need to predict product properties as a function of chemical process conditions. This is because the molecular composition is an optimal starting point for the prediction of mixture properties. Reactivity is an especially significant property that can be discerned given a molecule's (and its reaction environment's) structure. Other properties fall into performance and environmental classes. It is also timely to exploit the ability of molecule-based models to link atomic and process levels of chemistry. For example, molecule-based models can receive reactivity information as input from *ab initio* and/or semi-empirical quantum chemical calculations, use it in the solution of the kinetics model that provides global rates and selectivities, and in turn provide this kinetics information for a full process simulation. This is because the MBM is based on true chemical structure and not the lumped paradigm that has now outlived its usefulness.

Thus, the potential advantages of molecule-based modeling are clear. Less readily apparent, however, is that the development and operation of molecular models comes with a large requirement for model construction and solution time as well as reactivity information.

The challenge of building detailed kinetic models is due to the staggering complexity of the complex reaction mixtures. There will often be thousands of potential molecular and intermediate (e.g., ions or radicals) species. The sheer size of the thus-implied modeling problem engenders a conflict between the need for molecular detail and the formulation and solution of the model. Clearly, the use of the computer to not only solve but also formulate the model would be helpful in that it would allow the modeler to focus on the basic chemistry and rules of the model.

Our recent work has led to the development of such an automated capability to model development. It follows the four conceptual steps listed above. When the starting reactants are complex, the **(Step 1, CME)** simulation of feedstock structure casts the modeling problem in molecular terms. The model equations are then built **(Step 2, INGen, aka NetGen)** and coded **(Step 2, The Kinetic Model Editor, aka KME)** on the computer. Reactivity information is then organized in terms of quantitative linear free energy relationships **(Step 3, LFER's)**. Solution of this chemical reaction network, in the context of the chemical reactor **(KME)**, provides a prediction of the molecular composition, which is then organized into any desired commercially relevant outputs **(Step 4, CME)**.

Our research program at the **University of Delaware** and the Center of Research Excellence in Petroleum Refining & Petrochemicals at the **King Fahd University of Petroleum and Minerals** is aimed at the continued development of approaches and software tools for building molecule-based chemical kinetic models. Together, these products allow for the automated construction, editing, tuning and execution of kinetic models. It is important to note that the kinetic model, i.e., the set of mathematical equations, is fully separable from the KMT (e.g., KME, INGen, CME, etc.) that produced it.

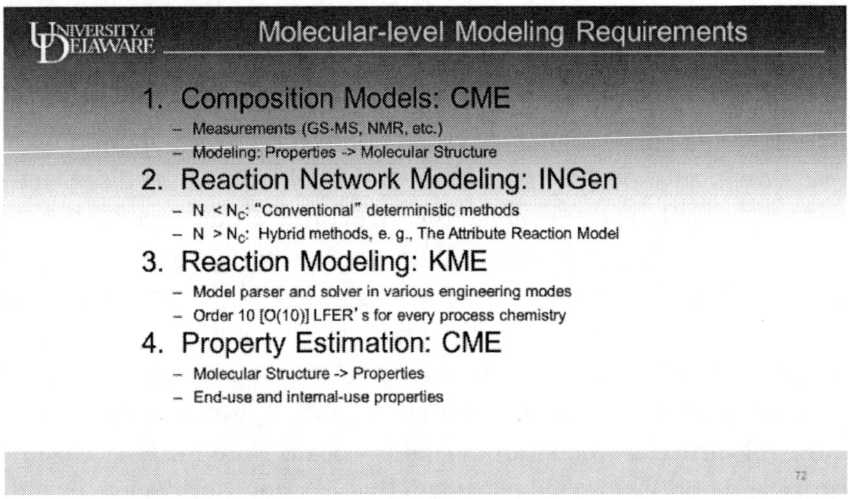

The Interactive Network Generator (INGen)

The models that will support the utilization of unconventional feedstocks in meeting the world's energy challenge will comprise tens of thousands of measurable molecules. This means that constructing the models automatically is not just an attractive option but an absolute requirement. The INGen tools were designed for that purpose.

The construction of the species' balance equations exploits the representation of a chemical species' atomic connectivity (molecule, intermediate, etc.) as a mathematical graph and the changes brought about by chemical reactions as the result of a simple matrix operation. A chemical species can be represented by a graph, G, where the set of vertices, V, and the set of edges, E, have a one-to-one correspondence with the atoms and bonds of the molecule, respectively. The connectivity information contained in the graph is stored in the computer in terms of the efficient adjacency structure, which is composed of a set of lists, one for each vertex. The list for a vertex v contains all the vertices connected to it. An equivalent but more mathematically tractable representation of the species is the bond and electron (BE) matrix. The diagonal elements, ii, of the BE matrix give the number of nonbonded valence electrons of atom i, and off-diagonal entries, ij, provide the connectivity and bond order of atoms i and j.

It is noteworthy that this BE matrix representation is also used in quantum chemical calculations, which facilitates the atomic-to-process integration noted above.

The BE matrix allows for chemical reaction to be a simple, dense matrix addition operation. The transformation R + B = E, i. e., the addition of the reaction matrix, R, to the reactant submatrix, B, yields a product submatrix, E, composed of the same atoms of the molecule with altered connectivity. The off-diagonal entries in the reaction matrix define the formation or breakage of bonds, and the diagonal elements specify the loss or gain of free valence electrons. When the reactive atoms come from two different molecules, as in bimolecular reactions, the i x i BE matrix of reactant 1 and the j x j BE matrix of reactant 2 are merged into a single (i +j) x (i + j) BE matrix from which the minimal representation, i. e., those atoms involved in the reaction, can be obtained.

Exhaustive application of these operations to not only the reactant(s) but also the thus-generated products produces the complete list of reactant-product relationships, i. e., the reaction network. Various subroutines assure species uniqueness, check connectivity and perform the geometric conditioning necessary to obtain species properties from semi-empirical quantum mechanical calculations for a first-pass estimate of the kinetics (see below)[1]. These software tools allow for the construction of the kinetic model balance equations, often totaling into the hundreds or thousands, in seconds.

The Kinetic Model Editor (KME)

After an initial and intensive use of INGen for the construction of the model equations, the model building process turns to other issues involved in the confrontation of the model predictions with experimental reality. Indeed, the bulk of the model building time is usually spent in activities subsequent to the mechanism generation, such as model tuning, editing and customization to a particular data set or set of process operating conditions. The Kinetic Model Editor (KME) focuses on the time consuming parts of model building: model tuning (parameter estimation), model editing, customization to a particular reactor type in context of a process chemistry, and so on.

KME provides a user-friendly environment for model development by taking either user-supplied or INGen/NetGen-generated reaction networks and automatically integrating them, along with the supporting programs and numerical solvers, into the computer model programs. The reaction mechanism editing and model rebuilding features are particularly useful since frequent revisions and modifications of the

---

[1] This is also the algorithmic link to the full power of the quantum chemical calculations.

reaction network and its underlying mechanisms are almost inevitable. With the spreadsheet-based graphical user interface, KME is not only the ideal model building and editing tool but also the perfect environment for model execution and simulation data analysis.

The Kinetic Model Editor was designed as a set of parsing tools to facilitate the conversion of reaction networks into mathematical equations written in C. An Excel-based user interface was used to automate the subsequent modeling procedures, including the model configurations, data entry, compilations and executions. The user interface not only provides a step-by-step guide to the model building, but also ensures the consistency and precision of the internal connections between different parts of the model application.

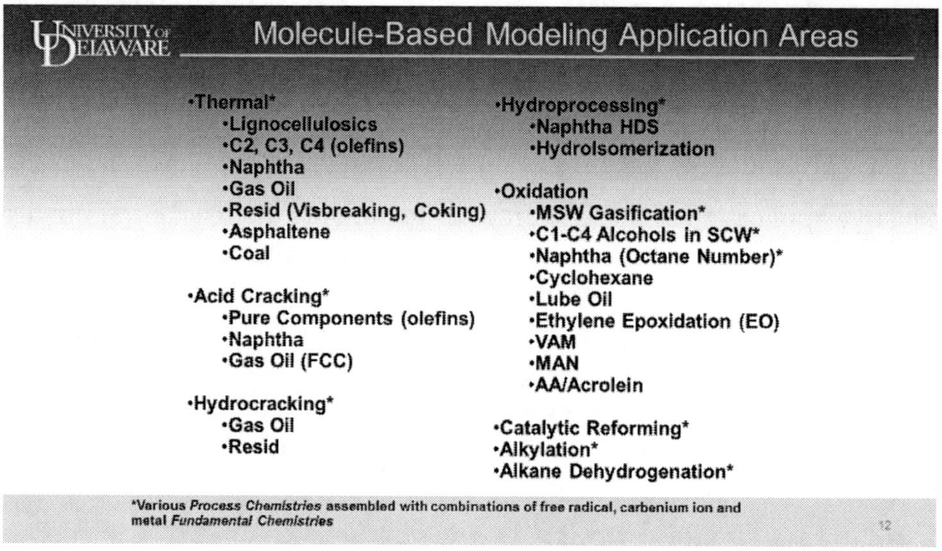

KME is capable of handling both homogeneous and heterogeneous reactions, detecting the existence of surface or intermediate species and identifying reversible and irreversible reactions.

CME

The Composition Modeling Editor (CME) is a front end to KME for modeling the composition of complex feeds. It converts bulk measurements (e.g., PONA, SARA, SimDis, H/C, etc.) into a molecular representation of the feedstock by optimizing a set of PONA/Carbon Number-oriented probability density functions (pdf's). At each iteration step during the optimization of these pdf's, CME transforms the molecular representation into a set of bulk properties. Thus, the application of CME to the final predictions of KME provides the product property capability of Step 4, above.

The Kinetic Parameters

Modeling complex mixtures at the molecular level always raises the question of the number and source of the rate constants. If the reaction model contains hundreds or thousands of species and reactions it will contain the same order of magnitude of rate constants. If these rate constants were to be derived from a purely experimental program the required database of reaction pathways and kinetics would be prohibitively enormous because of the practically impossible number of direct experiments required. The other approach of direct parameter estimation, where the model parameters are fit to experiments with real feeds, suffers because simultaneous regression of many parameters often leaves them individually devoid of chemical significance. Clearly methods must be employed to reduce the number of model parameters.

More careful scrutiny of the composition and reactions of complex mixtures suggests much of the complexity is statistical. Each species will generally fall into one of a handful of compound classes (e.g., paraffins, olefins, naphthenes, aromatics, alkylaromatics, etc.) and these in turn react in a manageable number of reaction families (β-scission, dehydrogenation, dealkylation, isomerization, etc.). Thus the complexity is really the reactions of sets of many similar compounds, where within each set compounds differ only in substituents. Essentially, differences in rate constants can be traced to substituent effects. This suggests that the use of linear free energy relationships (LFERs) as an organizational technique and correlational tool for assembling a substantial kinetic data base would be of great utility in quantitative modeling.

The use of LFERs in heterogeneous catalytic kinetic systems has been sparse because of the ambiguity concerning the mechanism of elementary steps and a possible rate determining step. However, our recent experimental work has shown how 512 hydrogenation and hydrocracking rate constants for the reactions of eight reactants and some 40 products at 7 temperatures over a single catalyst could be represented by 18 LFER-like structure/reactivity correlations. This represents about a 95% reduction in the number of parameters needed for modeling. Moreover, the number and values of model parameters would remain fixed as the number of reactants increased and/or the initial composition varied.

This work suggests that a handful of structure/reactivity correlations can indeed account for the reactivity of a wide range of hydrocarbon feed stocks, so long as the variations of molecules are within the experimentally studied reaction families. This limitation is not severe, and the development of feed stock-sensitive models can proceed readily. However, it is clear that our current work on the extension of the reaction family notion to include the concept of a "catalyst family" will be of great practical value.

Final Remarks

In summary, in the KFUPM-Saudi Aramco-UD collaborative program we are developing a set of software tools that allow chemical engineers and chemists to develop useful models rooted in fundamental concepts of organic, physical and computational chemistry. These tools will be illustrated using examples of kinetic models developed for refining and petrochemical applications.

Paper No. 2

Hydrogen production by steam reforming reactions over uniform Ni-Fe alloy nanoparticles

Keiichi Tomishige
Graduate School of Engineering, Tohoku University, Japan

Abstract

Production of hydrogen and synthesis gas is one of important technologies in the petroleum and chemical industry. The synthesis gas can be converted to useful liquid fuels and chemicals such as Fischer-Tropsch oil, methanol, and dimethyl ether. Hydrogen is practically produced from natural gas, and the produced hydrogen is mainly used for the ammonia synthesis. In particular, much attention has been paid to the hydrogen production relating to the fuel cell technology. Nickel is an effective component for the steam reforming reactions in terms of the catalytic activity and the catalyst cost, however, the coke deposition is one of the problems of the Ni catalysts, which is connected to the catalyst deactivation. In order to improve monometallic Ni catalysts, various kinds of modification of Ni catalysts have been attempted. Here, the alloying of Ni with Fe was applied to the catalyst development for the steam reforming reactions. However, the composition of Ni-Fe alloy particles is not so uniform on the catalysts prepared by usual co-impregnation methods. Uniform Ni–Fe alloy nanoparticles are successfully prepared by calcination and reduction of a hydrotalcite precursor. The nanocomposite structure consists of uniform Ni–Fe alloy nanoparticles and Mg(Ni, Fe, Al)O nanoparticles, and enables a high catalytic performance in the steam reforming of toluene, and this catalyst system has great potential for application in the steam reforming of other hydrocarbons and biomass tar.

Paper No. 3

Novel approach to overcome mass-transfer limitation in multiphase reactions with Taylor-Couette flow

Mohammad F Aljishi|*
Research and Development Center, Saudi Aramco, Dhahran 31311, Saudi Arabia
*Mohammad.aljishi@aramco.com

Abstract

Designing efficient reactors with optimum process intensification is of great importance to many industrial applications. Intensified processes, with higher mass transfer rates, can also lead to lower energy costs and capital costs, due to compact equipment size. Understanding mixing characteristics and influence of flow behavior on mass-transport is critical to enhancing performance in reactive flows, where the interaction between fluid dynamics and chemical reactions is important. This presentation explores the use of vortex flow in a continuous Taylor-Couette Reactor (TCR) in applications that require enhanced mixing and process intensification, as a unique approach to overcome the technical barrier in multiphase reactions. Specifically, the presentation assesses the effect of changing flow structure and mixing patterns in TCR on mass transport, for a range of multiphase flow reactions including crystal formation, particle functionalization and gas-liquid absorption.

As a model reactor system, where flow takes place in the annulus between two concentric cylinders (Fig.1), Taylor-Couette flow allows for examining hydrodynamic instabilities, while stimulating vortex motion, offering a highly active extensive interface for mass transfer, and multiphase mixing. Although the flow is steady and purely azimuthal for low angular cylindrical velocities, a series of hydrodynamic instabilities emerge during the gradual increase of inertial forces against viscous forces. Emerging vortices act to increase the surface area of the interface between different and partially unmixed materials, making Taylor Couette reactors desirable for enhancing reactions in multi-phase flow.

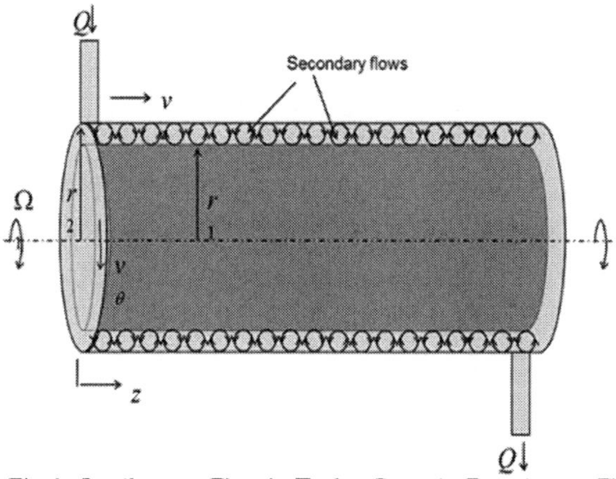

Fig.1. Continuous Flow in Taylor-Couette Reactor

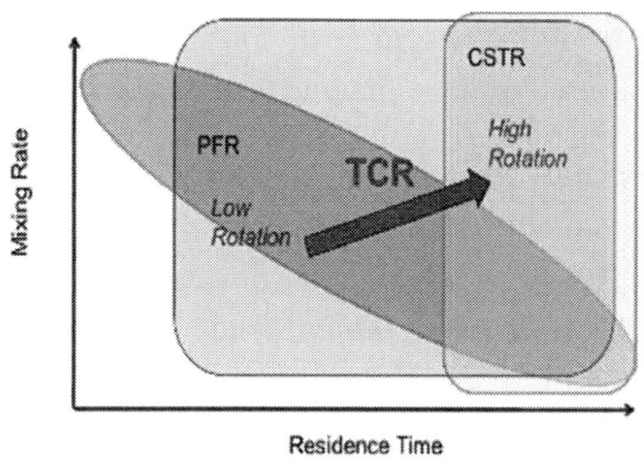

Fig.2. Mixing Rate vs. Residence Time in Various Reactors

Through the use of secondary flow regimes, the TCR provides high mass transfer rates and enhanced mixing properties, including a high surface-to-volume ratio, which makes it preferable for highly exothermal or endothermal reactions. In addition, narrow residence time distribution (reaching the behavior of plug-flow reactor) and homogeneous distribution of mixing intensity in the TCR volume (compared to a continuous-stirred-tank reactor), combined with mild local shear rates makes TCR an attractive reactor for a range of chemical reactions (see Fig.2). While the TCR may be used for any solid-liquid-gas reaction, the reactor is particularly useful for difficult feed streams such as those containing high viscosity (hydrocarbon emulsions), as well as suspended solids (with applications relating to precipitation, catalyst synthesis, coating & surface functionalization).

Assessment of reactor performance as a function of operational and design parameters is critical in demonstrating the utility of the TCR for multi-phase flow reactions. In this presentation we explore the effect of change in flow structure during the onset of primary instability in Taylor Couette flow, which results in the transition from purely azimuthal laminar flow to the emergence of Taylor vortices, on mass transport with regards to a couple of model chemical reactions.

The first part of this study focuses on examining the homogeneous reaction of barium sulfate precipitation in Taylor–Couette crystallizers. Experiments involving precipitation of barium sulfate under laminar flow (before the onset of instability) and vortex flow (after the onset of primary instability) at three different axial flow rates revealed that the vortex motion plays a critical role in fine-tuning crystal properties such as internal crystal structure, particle microstructure and morphology. In particular, we observed a reduction in the average crystal size and a much narrower crystal size distribution (CSD) after the emergence of Taylor vortex flow.

The second part of this study examines the influence of vortex flow in tailoring the coating layer during the heterogeneous reaction involving the salinization of metallic filler particles with methacryloxypropyltrimethoxysilane (MPS). The surface of aluminum fillers was successfully modified with specific ($H_2C=C$) functionality offering a more uniform degree of coating, lower degree of aggregation, smoother surface and higher coupling efficiency after the emergence of Taylor vortices.

Based on results from these two studies, the variety of flow patterns in a three-phase system, reacting gaseous carbon dioxide (CO_2) with aqueous calcium hydroxide ($Ca(OH)_2$) to precipitate solid calcium carbonate crystals ($CaCO_3$), were investigated in the TCR to assess the influence of the corresponding flow dynamics, CO_2 gas bubble size and interfacial area on the properties of the resulting calcium carbonate crystals. The study included experimental characterization of flow patterns by assessing gas bubble dispersion and bubble size, to identify flow-patterns with highest interfacial area. The ideal mixing flow regimes identified from these experiments were used for calcium carbonate precipitation. It was found that identifying the emergence of the "ring flow" regime is critical to optimizing the interfacial area and to fine-tuning crystal properties such as internal crystal structure, particle size and morphology.

References:

1. Aljishi, M. F.; Ruo, A.-C.; Park, J. H.; Nasser, B.; Kim, W.-S.; Joo, Y. L. Effect of flow structure at the onset of instability on barium sulfate precipitation in Taylor–Couette crystallizers. *Journal of Crystal Growth* 2013, *373,* 20–31.
2. S. H. Kang, S. G. Lee, W. M. Jung, M. C. Kim, W.-S. Kim, C. K. Choi and R. S. Feigelson, "Effect of Taylor Vortices on Calcium Carbonate Crystallization by Gas-Liquid Reaction," *Journal of Crystal Growth,* no. 254, pp. 196-205, 2003.
3. Aljishi, M.F.; Joo, Y.L. Influencing surface functionalization of aluminum fillers with acrylic-monomers through the onset of instability in Taylor Couette flow. Conference Proceedings of 250th American Chemical Society National Meeting & Exposition, 2015.

Paper No. 4

Supported Keggin-type heteropolyacid catalysts for isobutene oligomerization and methacrolein oxidation

Yuich Kamiya
Faculty of Environmental Earth Science, Hokkaido University, Japan

Abstract

Keggin-type heteropolyacids (HPA) including H4SiW12O40 and H3PMo12O40 have Brønsted acid sites with acid strength stronger than concentrated H2SO4 and oxidizing ability and thus show high catalytic activities for wide variety of acid-catalyzed reactions and selective oxidations. However, solid HPA do not often work well as a catalyst because of their small surface areas. To overcome this drawback, supporting HPA on a high surface area carrier like SiO2 is one of the effective methods to increase a number of HPA exposed on surface, consequently to enhance the catalytic activities. In this presentation, our studies on changes in structure, and acid and oxidation properties by supporting HPA on SiO2 are introduced. In addition, two examples of catalytic reactions including preferential oligomerization of isobutene in butenes mixture over H4SiW12O40/SiO2 and selective gas-phase oxidation of methacrolein over H4PMo11VO40/SiO2 are described.

Paper No. 5

Hydroprocessing catalysts for heavy oil upgrading

Lianhui Ding
Saudi Aramco R&DC

Abstract

Recent technologies in heavy oil hydroprocessing upgrading and studies on new hydrocracking catalysts application in heavy oil processing are presented. Due to the thermodynamic limitation, polyaromatics in heavy oil cannot effectively be converted under conventional hydrotreating conditions and standard catalysts. Although the hydrocracking catalysts can improve the ring saturation and opening of some of the aromatics compounds, the opening sizes of the zeolites remain too small to allow larger aromatic molecules to diffuse into the active sites located inside the zeolite. A decrease of zeolite crystals to nanometer scale can lead to shorter diffusional paths of the reactant and product molecules inside the pores and consequently to a reduction or elimination of undesired diffusional limitations of the reaction rate. Data on nano-sized zeolite beta that was successfully synthesized and evaluated for heavy oil hydroprocessing are presented.

Extended Abstract

Hydrocracking process is one of the most important processes in modern refineries, due to its high versatile in processing different feedstocks, producing the good quality with high liquid yields, especially at times when crude oil properties are getting deteriorated yearly, and more environmental friendly products are demanded. Hydrocracking catalysts play a crucial role in the process, because the catalysts determinate the product selectivity, hydrogen consumption, run length, and economics. Hydrocracking catalysts are bi-functional: cracking function contributed by the acidic sites such as amorphous oxides and/or zeolites, and hydrogenation function by the metals such as Co-Mo, Ni-Mo, or Ni-W.

Zeolite beta was proved to be one of the most suitable acid components for use in heavy oil hydroconversion due to its appropriate pore structure, acidity, and hydrothermal stability. However its interconnected 12-membered ring channels make it difficult for large molecules present in heavy oil fraction to diffuse to the inner surface where most of the reaction sites are located. An effective way to reduce or eliminate the undesired diffusion limitation it to reduce the crystal sizes of the zeolites. When the particle sizes are reduced to nano-size, substantial changes in the physicochemical properties can be obtained. Nano-sized zeolite beta crystals were reported to exhibit higher catalyst activity, a lower rate of catalyst deactivation, and higher product quality than conventional microcrystalline beta materials.

This presentation presents the hydrothermal synthesis method of the nano-sized zeolite beta in the TEAOH-SiO-Al-H2O system in the absence of alkali cations. The effect of agitation under various TEAOH/SiO2 molar ratios on the crystallization kinetics, particle sizes, particle size distribution, and framework Si/Al ratio were studied. The synthesized zeolites were characterized by XRD, TEM, N2 adsorption. In order to compare the performances of the nano-sized and conventional micro-sized zeolite beta, a hydroprocessing test was carried out in an autoclave with light cycle oil (LCO) from FCC process.

The experiment results are summarized in the table below. The results show that the effect of the agitation on the induction time, the particle size, and the framework Si/Al ratio was pronounced at high TEAOH/SiO2 ratio than at a low TEAOH/SiO2 ratio. Under agitation, the induction time, the particle sizes, and framework Si/Al ratio greatly decreased with the increase of the TEAOH/SiO2 ratio.

	State	TEAOH/SiO_2	Size(nm)	Framework Si/Al ratio	Pore volume(ml/g)
S-2	Static (in oven))	0.21	480	32.4	0.380
S-4		0.4	75	37.6	0.801
S-6		0.6	100	38.8	0.870
A-2	Agitation (in oil bath)	0.21	500	47.6	0.392
A-4		0.4	110	38.5	0.883
A-6		0.6	30	31.6	0.896

The LCO hydroprocessing experiment results are shown in the figure below. The results show that the hydrodenitrogenation (HDN), hydrodesulfurization (HDS), hydrodearomatization (HDA) activities, and liquid yields of the catalyst with the nano-sized zeolite were significantly enhanced compared with the conventional micro-sized catalyst.

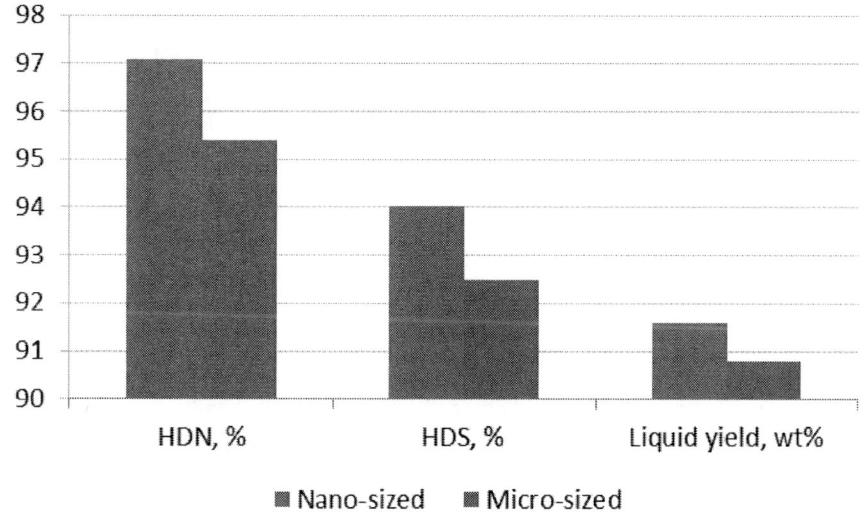

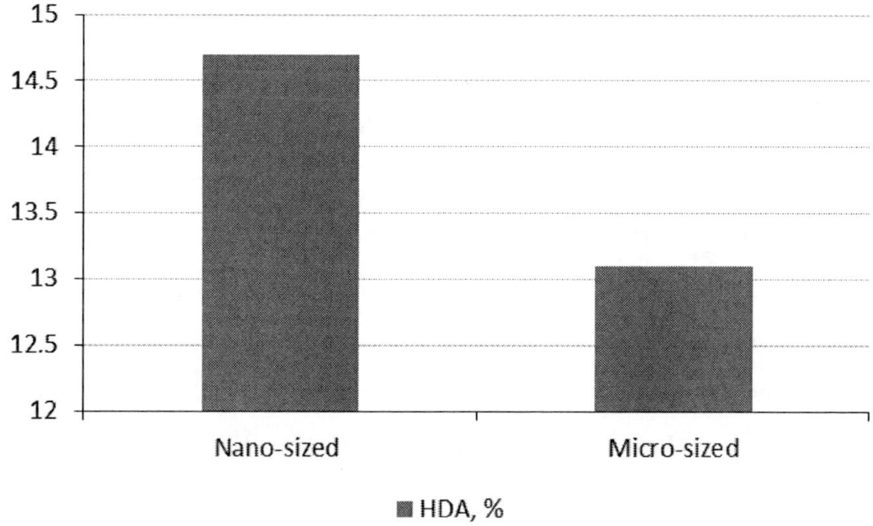

(The feedstock: LCO with aromatics: 87wt%; Catalysts: Ni-W/zeolite beta+alumina; conditions: 375°, 68bar, $0.5h^{-1}$)

Paper No. 6

SPH (Slurry Phase Hydrocracking) Process as a cutting-edge upgrader-for upcoming diesel demand in not-too-distant future

Daisuke Oka
Chiyoda Corporation, Japan

Abstract

Slurry phase hydrocracking is deemed as an ultimate process for upgrading heavy oils, realizing zero worthless fuel oil refining and maximizing transportation fuel production. This enables refiners to meet a future oil market demand by conversion of ultra heavy oils, increasing production of transportation fuels, as well as meeting demanding new environmental constraints. To make it realize, Kobe Steel, Ltd. And Chiyoda Corporation has been developing an innovative process, named the KOBELCO SPH process, which applies our past slurry process experience, well-established brown coal liquefaction technology, to it. This can be applied as an upgrading technology either for refineries or upstream processing. In the presentation, the outline, features, applications and advantages of the KOBELCO SPH process will be demonstrated. The process consists of three general steps: hydrocracking, where the ultra heavy oils are hydrocracked in slurry bed reactors with high-active and inexpensive dispersed natural limonite ore as catalyst, followed by in-line hydrotreating and finally solid-liquid separation. It can convert any type of ultra heavy oil into high-yield distillate products under relatively mild conditions without any plugging problem in the reactor. The in-line hydrotreating step, in which metals and asphaltene are removed, produce clean products which meet the current transportation fuels standards (e.g. Euro V).

Paper No. 7

CAN-15 catalyst- data review

Vinod Ramaseshan, Saudi Aramco P&CSD
and
Bandar Solami, Saudi Aramco R&DC

Abstract

Saudi Aramco Riyadh Refinery operates a Single Stage Recycle Hydrocracking Unit operating at a high conversion per pass. The feed to the unit is a mixture of straight run Vacuum Gasoil and De-Mineralized Oil from a Solvent Deasphalting Unit. The uniqueness of the feed is that the DMO is extremely heavy with its 90% point above 700°C. Saudi Aramco have joint developed with JGC C&C a tailored catalyst CAN-15 for use in this service. This paper will essentially talk about the first commercial application of the CAN-15 catalyst. This catalyst has been partially loaded along with essentially regenerated catalyst (to the tune of 80%) at Riyadh Refinery and has been successfully operated for over 2 years of operation. The paper will talk briefly of the Hydrocracking Unit, CAN-15 catalyst, Pilot Plant testing and optimal catalyst loading strategy used to mitigate the risks of first commercial usage and the operating data for the current cycle. CAN-15 catalyst has been successfully commercialized and the yield structure for the current cycle exceeds expectations.

Paper No. 8

Oil refining in a CO_2 constrained world with implications for gasoline and diesel fuels

Amir Abdul Manan
Saudi Aramco R&DC

Abstract

Diesel compression ignition engines are known to be more efficient, and therefore less CO_2 intensive, than their gasoline spark-ignition counterpart. This has been reflected in many Well-to-Wheels (WtW) studies to date, some of which underpin regulatory measures that favor dieselization. In this study we explore the CO_2 emissions benefits of diesel vs gasoline from a refining standpoint. We utilized refinery Linear Programming (LP) models to represent the global refining industry, and imposed a price on refinery CO_2 in order to optimize refinery productions under a CO_2 constrained world. Contrary to popular belief, we show that in a world that internalizes the cost of CO_2 emissions, refineries would favor gasoline over diesel production. This highlights that the global shift towards transport dieselization can lead to greater refinery CO2 emissions. We propose that an optimum fuel and engine combination is the use of a gasoline-like fuel in an efficient compression-ignition engine.

Paper No. 9

Development of the efficient process for producing BTX from fuel oil

Masahide Kobayashi
JX Nippon Oil & Energy Corporation, Japan

Abstract

The demand for fuel oil is gradually decreasing in Japan. Meanwhile, global demand for petrochemicals is annually increasing. Therefore, technology of converting fuel oil into petrochemicals is strongly required. We have developed efficient processes for producing BTX (Benzene, Toluene and Xylene) from LCO (Light Cycle Oil from FCC units) which is widely used as a fuel source to heavy oil and from light naphtha which is used as a fuel source to gasoline. FCA process is efficient in producing BTX from LCO in the absence of hydrogen gas. LCO is composed of compounds which have various molecular structures. The important function of FCA process is to optimize reactions of each structure, so we focused on comprehending the components of LCO in detail. Catalyst deactivation due to coke formation is a major issue of FCA process. To solve this issue, we developed fluid-bed reaction system to regenerate catalysts and to enable continuous reactions. Furthermore, we found partial hydrogenation of LCO as a pretreatment is an effective method to increase BTX yield. Z-FORMING process is efficient in producing BTX from LPG and light naphtha. We achieved producing BTX in a high yield by improved catalyst containing zeolite.

Paper No. 10

Selective dehydrocyclization reaction of n-paraffins using zeolite-oxide composite supported NiMo-based catalysts

Atsushi Ishihara, Kosuke Oda, Akihiro Itoh, Hiroyuki Nasu, Tadanori Hashimoto
Division of Chemistry for Materials, Graduate School of Engineering, Mie University, Japan

Abstrac

If dehydrocyclization of n-paraffins form aromatics, one of reactions occurring in the catalytic reforming, will proceed selectively, straight-run naphtha, which mainly consists of n-paraffins, can be regarded as a good hydrogen carrier. Although a Pt/Al2O3 catalyst with Cl ion is usually used for the reforming reaction, Pt is expensive noble metal and further it is difficult for this catalyst to make the selective dehydrocyclization of n-paraffins. Therefore the alternatives to Pt are strongly required. In this study we investigated the preparation of zeolite-alumina composite-supported Ni-Mo catalysts and their reactivity for the selective dehydrocyclization of n-heptane, a model compound of naphtha. It was found that ZSM-5-alumina composite-supported NiMo catalysts were very active for the selective dehydrocyclization of n-heptane to form hydrogen and aromatics. The conversion for NiMo/ZSM(24)60A increased with increasing temperature and reached 91 % at 500 oC. With the addition of Pt, PtNiMo/ZSM(24)60A showed the highest conversion (98%) and toluene selectivity at 500°C. NiMo/ZSM(24)60Ti showed the highest conversion and toluene selectivity at 400 and 450°C. The deactivation was not observed for these catalysts with the elapse of time at 500oC. On the other hand, NiMo/ZSM(24)60Mg showed the lowest conversion and toluene selectivity.

Paper No. 11

Anode-grade coke from Arabian crudes

Bandar Solami
Saudi Aramco R&DC

In recent years, Saudi Arabia had built its first Aluminum smelting refinery. Anode-grade coke is heavily used for aluminum production. This study investigated the technical feasibility of producing anode-grade coke from Arab Heavy (AH), Arab Light (AL), and Arab Heavy/Light 50/50 blend (BL) crudes. The crudes were first distilled to produce a vacuum residue (VR) with a boiling point higher than 1050 °F. The VR is then hydrotreated to reduce the sulfur and metals contents via an ebullated bed technology pilot plant using two reactor configurations: with two reactors (2-stage) and with three reactors (3-stage). The liquid products of the treatment process were further distilled to produce a treated 1050+ °F VR. The treated VRs were fed to a delayed coking pilot plant, applying two different operating conditions (Coke 1 and Coke 2), for the production of green anode-grade coke, which was calcined and converted into anode-grade coke. The study targeted an anode-grade coke with a sulfur content of 3.5 wt. %, vanadium content of 300 ppm, and nickel content of 250 ppm maximum. It is shown that VR from all the three crudes could be hydrotreated to sulfur levels of 1.5-2.0 wt. % and vanadium and nickel contents to less than 25 ppm each. Using the treated VRs in delayed coker, the production of anode-grade coke, meeting all specifications, was proven.

Proceedings 26th Saudi Japan Annual Symposium
Technology in Petroleum Refining & Petrochemicals
KFUPM Dhahran, Saudi Arabia Nov. 7-8, 2016

Paper No. 12

Active and stable catalyst for 1-butene cracking to propylene

Amr Abdalla, Palani Arudra, Sulaiman S. Al-Khattaf*

Center of Research Excellence in Petroleum Refining and Petrochemicals, King Fahd University of Petroleum & Minerals, Dhahran 31261, Saudi Arabia.

** Corresponding author. Tel.: E-mail address: skhattaf@kfupm.edu.sa*

Abstract

Core-shell silicalite composite material was synthesized using ZSM-5 as parent material by hydrothermal method. The synthesized material was characterized using various physiochemical methods such as XRD, TPD, N_2 adsorption-desorption and SEM analysis. Catalytic cracking of 1-butene was carried out using core-shell silicalite composite. The propylene yield (P/E ratio of 3.0) was higher for core-shell silicalite composite as compared to parent ZSM-5 catalyst. The higher catalytic activity of core-shell silicalite composite is related to effective control of acid sites, which reduces the hydrogen transfer reactions leads to alkanes and aromatics. The core-shell silicalite composite showed excellent stability over 50 hours of the 1-butene stream.

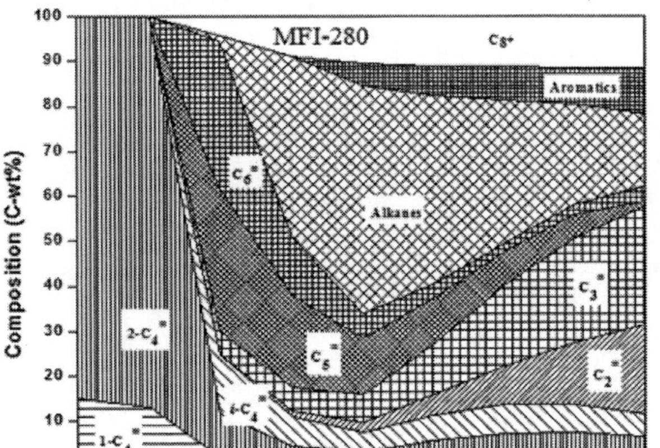

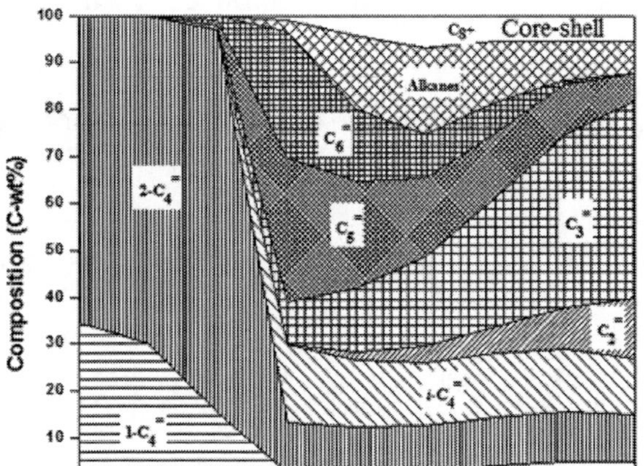

Paper No. 13

Highly selective zeolite based catalysts for naphtha catalytic cracking to ethylene and propylene

Nabil Al-Yassir [†a], Khalid Al-Majnouni [a], Ahmed Al-Zenaidi [a], Wojciech Supronowicz [a], Ahmad Alshehri [a], Naif Al-Dalaan,[a]
[(a)] Olefins, Industrial Gases & Aromatics, SABIC Technology and Innovation, Saudi Basic Industries Corporation (SABIC), Riyadh 11551, Saudi Arabia.

Sulaiman Al-Khattaf [b], Abdullah M. Aitani [b], M. A. Bari Siddiqui [b]
[(b)] Center of Research Excellence in Petroleum Refining & Petrochemicals, King Fahd University of Petroleum & Minerals, Dhahran 31261, Eastern, Saudi Arabia.

† Corresponding Author (Email: alyassern@sabic.com , Tel: + 966114999820)

Abstract

The demand on propylene has recently increased with a projection to reach more than 100 MT annually by 2016. Propylene is mainly produced by steam cracking (57%), refinery FCC (30%) and on-purpose production technologies such propane dehydrogenation and methanol (13%). However, with the increase in the gap between supply and demand, the on-purpose technologies contribution will reach 30% by 2019. SABIC in collaboration with KFUPM (Centre for Refining and Petrochemical) has developed several novel zeolite based catalysts that gave high yield of propylene and excellent stability. The novelty of these zeolites comes from the generation of hierarchical pore structures and controlling the extent of their formation as well as controlling the density of strong acid sites. The hierarchical structure of the catalysts provides faster mass transfer and higher accessibility to the active sites and prevents secondary reactions such as hydrogen transfer that consume olefins. The novelty of these zeolites has been demonstrated in lab scale with substantial increase in the yield of ethylene and propylene, when compared to conventional pristine zeolites.

Paper No. 14

A light olefin selective FCC catalyst based on Ti,Zr-inserted USY zeolite

Yaming Jin, Omer R. Koseoglu
Saudi Aramco R&DC

and

Seiji Arakawa, Kenji Nita
JGC C&C

Abstract

Following the successful commercial deployment of the CAN-15 hydrocracking catalyst at Saudi Aramco's Riyadh Refinery, the collaborating teams from Aramco R&DC and Japanese JGC C&C developed a new light olefin selective FCC catalyst based on a newly developed USY zeolite. The catalyst, labelled as CAN-FCC3, was tested in a lab scale ACE reactor for converting hydrocracking bottom (HCB) feedstock to light olefins, especially propylene and butenes, at the FCC Laboratory of KFUPM under a technical laboratory services agreement. Two sets of parallel tests on CAN-FCC3 were administered: (1) cracking catalyst only mode in comparison with a JGC C&C's standard FCC catalyst and (2) cracking catalyst with olefin additive mode in comparison with a top-tier olefin-selective FCC catalyst from another catalyst vendor.

The cracking activity of CAN-FCC3 catalyst was found to be similar to that of JGC C&C's standard FCC catalyst, but CAN-FCC3 was more selective in producing naphtha range olefins from the HCB feed. When a JGC C&C's proprietary olefin additive was included in the ACE testing at 10 wt.% of the catalyst load of CAN-FCC3, the combined light olefins ($C_2^=$-$C_4^=$) yield reached 54% (26.7% - propylene alone) in comparison with 40.6% (17.7% - propylene alone) of the test run with the top-tier olefin-selective vendor catalyst, which contains 10 wt.% olefin additive in the catalyst formulation.

Paper No. 15

An integrated study on the production of light olefins from catalytic cracking of Arab crude oil fractions in different reactor systems

Qi Xu
Saudi Aramco R&DC

Abstract

Light olefins, such as ethylene and propylene, are important building blocks for a wide range of important chemicals in the petrochemical industry. One of the promising technologies with the objective of maximizing propylene production is the high severity fluid catalytic cracking (HS-FCC) process, where the catalyst is fluidized, and takes part in reactions and the regeneration cycle continuously under high severity conditions. Hence, this study utilizes the concept of the high severity down flow reactors to assess the production of light olefins from various crude oil fractions through catalytic cracking reactions. Three crude oil fractions were tested in this study where the experiments were carried out in a microactivity test unit (MAT), a microdowner unit (MD), and a pilot plant unit, with a focus on intrinsic difference in the olefin production due to characteristic features of each reactor system. The range of the catalyst to oil ratio (C/O) in the MAT unit is lower in comparison with that in the MD and the pilot plant units, while the total feed conversion and olefin yields in the MAT unit is generally higher than those in the other two systems. Since the feed conversion differs in all three reactor systems, the comparison of the production yield distributions was carried out not based on the same conversion, but rather based on various reaction severity conditions (C/O). The differences in the experimental results between the reactor systems were explained in terms of reaction contact time and reaction mechanisms. Then the experimental data were analyzed using linear regression and fitted to a proposed mathematical model. The coefficients of determination (R2) for each set of correlations are calculated. The resulting R2 values (>0.95) indicate good correlation between the predicted yields and the actual yields for all three crude oil fractions. This pragmatic analysis provided an integrated view of the limitations and advantages of each reactor system, which helped design improved catalyst formulations for the studied feedstocks to enhance the production of light olefins.

Paper No. 16

Study of the nano-structured zeolite for hydrocarbon cracking

Emad N. Al-Shafei [a], Oki Muraza [b], Ki-Hyouk Choi [a], Zain H. Yamani [b], Mohammed A. Sanhoob [b], Mohamed A. Abdullah [a], Ashok K. Punetha [a] and Ali S. Alnaser [a]

a) Research and Development Center, Saudi Aramco, Dhahran 31311, Saudi Arabia
b) Center of Excellence in Nanotechnology and Chemical Engineering, King Fahd University of Petroleum & Minerals (KFUPM), Dhahran 31261, Saudi Arabia.

Abstract

Structured zeolites are the potential catalysts for the refinery and petrochemical industry applications. This study aims to develop and investigate the nano-structured zeolite for the hydrocarbon conversion with better hydrothermal stability [1-2]. There are several challenges that limited utilizing zeolites in a wider range for cracking applications, including selectivity expectation, stability requirement and high coke formation. In order to overcome these intrinsic limitations, zeolite crystal size was reduced to the nanoscale to shorten the diffusion path as compared with the conventional microsized zeolite, to improve the accessibility of larger molecules on mesoporous surface and to accelerate the reaction rate. Although there have been many zeolite applications based on their microporous structures, this study explores the effect of synthesis parameters on the catalyst activity and the hydrothermal stability of nanoscale MFI zeolites. To develop the hydrophobic zeolites, the fluoride media, and rare earth metals were used as the hydrophobic agent. The synthesized nanocrystal zeolite catalysts were characterized by ammonia temperature programmed desorption (NH_3-TPD), XRD, NMR, and N_2 physisorption. The hydrocarbon cracking catalytic testing was carried out by using fixed bed reactor at the space velocity of 2-8 h^{-1} to investigate the conversion, selectivity, reaction stream stability and coking rate with an emphasis to compare the nanostructure and microstructure zeolites. Also, the presentation will discuss the reaction mechanism of shape selectivity of nanocrystal zeolites at moderate temperature and effect of steam addition to the zeolite stability.

Introduction

There have been an increasing interest in the utilization of heavy oil residue feedstock as a source of fuel and other value-added chemicals. Zeolites are widely applied in upgrading oil due to their catalytic performance and economic viability [1-3]. To the objective is to avoid the undesirable side reactions during the heavy hydrocarbon upgrading,poor stability of zeolites and high degree of coking of catalysts. Zeolites are well studied and prepared to be stable in the presence of steam environments, yet they are vulnerable to the attacks by liquid water molecules at aqueous elevated temperatures environment where they are reported to lose their crystallinity and catalytic activity [4]. On the other hand, this vulnerability of zeolite framework toward hydrothermal treatments in steam can open new doors to effectively modify and even to synthesize new zeolites [5-8]. Several factors affect the hydrothermal stability of zeolites such as zeolite framework shape[9], pore opening size[9], Si/Al ratio[10], type of cation [9], crystal size [11] and the presence of structural defects [12-13]. The pH and temperature of the aqueous solution highly determine the degradation mechanism of zeolites [14]. In this work, the hydrothermal stability of nano size one-dimensional and three-dimensional pore zeolite ZSM-22 and ZSM-5 with different Si/Al ratios was investigated in hot water environments under different temperatures and different exposure times.

Discussion

ZSM-22 zeolites, as can be observed from the XRD patterns in Figure 1, were synthesized successfully with nominal Si/Al ratios in the range of 30 to 100. The main representative peaks of ZSM-22 zeolite phase as reported elsewhere [15-17] were found in agreement with the synthesized samples. However, traces of cristobalite impurities at 2Θ = 21.8 were observed in the samples with Si/Al higher than 30 (Si/Al = 46, 80 and 100). ZSM-22 zeolites with nominal Si/Al of 46 were tested in the presence of hot water for different treatment times and different temperatures to evaluate their hydrothermal stabilities with emphasis on crystalline structure, textural properties, zeolite acidity, Si and Al contents as well as their locations in the zeolite structure.

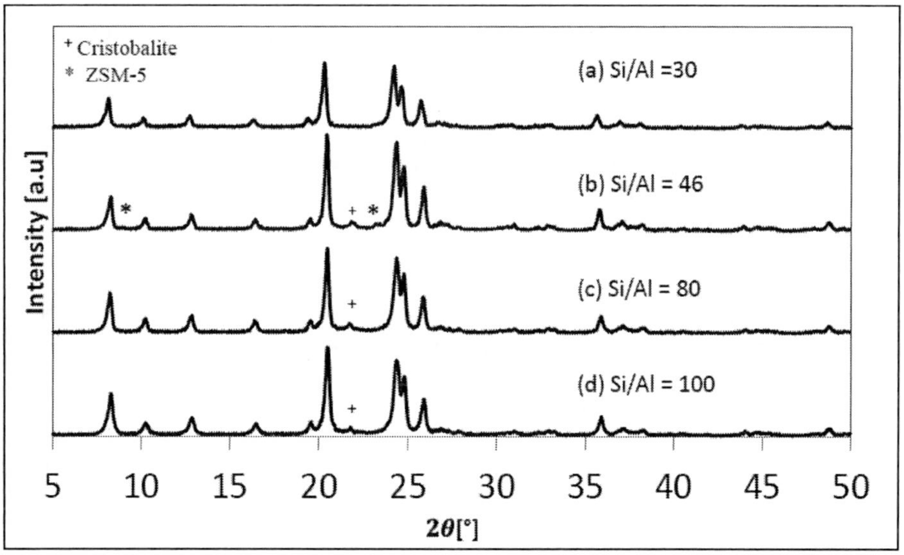

Figure S1. XRD patterns and crystallinities of H-ZSM-22 zeolites different Si/Al ratios: (a) 30, (b) 46, (c) 80 and (d) 100[1].

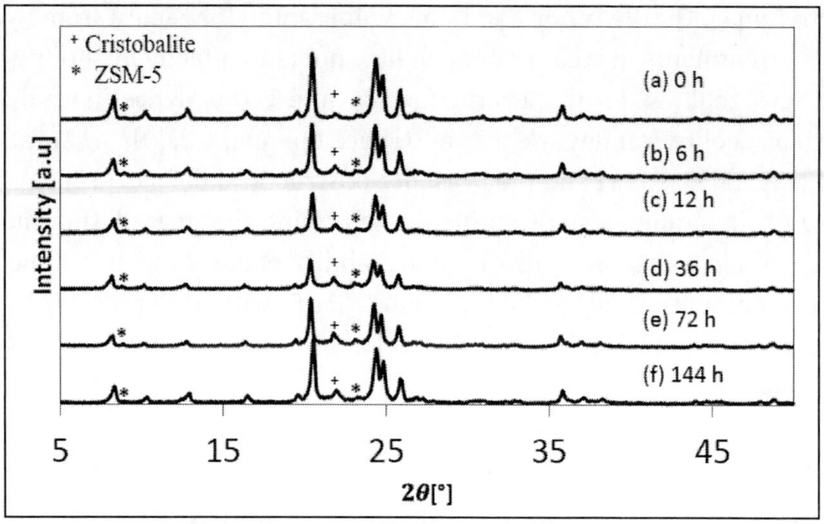

Figure 1. XRD patterns of H-ZSM-22 zeolites with Si/Al of 46 treated in hot water at 250 °C and different treatment times: (a) 0, (b) 6, (c) 12, (d) 36 h, (e) 72, (f) 144 h [1].

Conclusions

Regardless of the hydrothermal treatment condition, the crystalline structure of all tested MFI samples remained intact and did not completely collapse into amorphous products after the treatments in hot water. However, changes in the textural properties, relative crystallinity, and zeolitic phase were observed. Overall, MFI zeolite showed fair hydrothermal stability for a long time of 72 h at temperatures lower than 250 °C.

Reference

1. Anas K. Jamil, Oki Muraza, Ryota Osuga, Emad N. Shafei, Ki-Hyouk Choi, Zain H. Yamani, Ali Somali, and Toshiyuki Yokoi, Hydrothermal stability of one-dimensional pore ZSM-22 zeolite in hot water, Journal of Physical Chemistry C (2016) 120 (40)

2. Idris A. Bakare, Oki Muraza, Teguh Kurniawan, Zain H. Yamani, Emad N. Shafei, Ashok K. Punetha, Ki-Hyouk Choi, Toshiyuki Yokoi, Hydrothermal stability of MTT zeolite in hot water: The role of La and Ce, Microporous and Mesoporous Materials 233 (2016) 93-101.

3. Ennaert, T.; Van Aelst, J.; Dijkmans, J.; De Clercq, R.; Schutyser, W.; Dusselier, M.; Verboekend, D.; Sels, B. F., Potential and Challenges of Zeolite Chemistry in the Catalytic Conversion of Biomass. *Chemical Society Reviews* **2016**, *45*, 584-611.

4. Zapata, P. A.; Faria, J.; Ruiz, M. P.; Jentoft, R. E.; Resasco, D. E., Hydrophobic Zeolites for Biofuel Upgrading Reactions at the Liquid–Liquid Interface in Water/Oil Emulsions. *Journal of the American Chemical Society* **2012**, *134*, 8570-8578.

5. Eliášová, P.; Opanasenko, M.; Wheatley, P. S.; Shamzhy, M.; Mazur, M.; Nachtigall, P.; Roth, W. J.; Morris, R. E.; Čejka, J., The Ador Mechanism for the Synthesis of New Zeolites. *Chemical Society Reviews* **2015**, *44*, 7177-7206.

6. Roth, W. J.; Shvets, O. V.; Shamzhy, M.; Chlubná, P.; Kubů, M.; Nachtigall, P.; Čejka, J. i., Postsynthesis Transformation of Three-Dimensional Framework into a Lamellar Zeolite with Modifiable Architecture. *Journal of the American Chemical Society* **2011**, *133*, 6130-6133.

7. Morris, R. E.; Čejka, J., Exploiting Chemically Selective Weakness in Solids as a Route to New Porous Materials. *Nature chemistry* **2015**, *7*, 381-388.

8. Mazur, M.; Wheatley, P. S.; Navarro, M.; Roth, W. J.; Položij, M.; Mayoral, A.; Eliášová, P.; Nachtigall, P.; Čejka, J.; Morris, R. E., Synthesis of 'Unfeasible'zeolites. *Nature chemistry* **2015**.

9. Roth, W. J.; Nachtigall, P.; Morris, R. E.; Wheatley, P. S.; Seymour, V. R.; Ashbrook, S. E.; Chlubná, P.; Grajciar, L.; Položij, M.; Zukal, A., A Family of Zeolites with Controlled Pore Size Prepared Using a Top-Down Method. *Nature chemistry* **2013**, *5*, 628-633.

10. Lutz, W.; Toufar, H.; Kurzhals, R.; Suckow, M., Investigation and Modeling of the Hydrothermal Stability of Technically Relevant Zeolites. *Adsorption* **2005**, *11*, 405-413.

11. Lutz, W.; Zibrowius, B.; Löffler, E., Hydrothermal and Alkaline Stability of High-Silica Y-Type Zeolites in Dependence on the Dealumination Procedure. In *Studies in Surface Science and Catalysis*, J. Weitkamp, H. G. K. H. P.; Hölderich, W., Eds. Elsevier: 1994; Vol. Volume 84, pp 1005-1012.

12. Ding, L.; Zheng, Y.; Hong, Y.; Ring, Z., Effect of Particle Size on the Hydrothermal Stability of Zeolite Beta. *Microporous and Mesoporous Materials* **2007**, *101*, 432-439.

13. Prodinger, S.; Derewinski, M. A.; Vjunov, A.; Burton, S. D.; Arslan, I.; Lercher, J. A., Improving Stability of Zeolites in Aqueous Phase Via Selective Removal of Structural Defects. *Journal of the American Chemical Society* **2016**, *138*, 4408-4415.

14. Zhang, L.; Chen, K.; Chen, B.; White, J. L.; Resasco, D. E., Factors That Determine Zeolite Stability in Hot Liquid Water. *Journal of the American Chemical Society* **2015**, *137*, 11810-11819.

15. Xiong, H.; Pham, H. N.; Datye, A. K., Hydrothermally Stable Heterogeneous Catalysts for conversion of Biorenewables. *Green Chemistry* **2014**, *16*, 4627-4643.

16. Masih, D.; Kobayashi, T.; Baba, T., Hydrothermal Synthesis of Pure ZSM-22 under Mild Conditions. *Chemical Communications* **2007**, 3303-3305.

17. Valyocsik, E. W., Synthesis of Zeolite ZSM-22. Google Patents: 1990.

Paper No. 17

Zeolite supported structured catalysts for methanol to propylene conversion

Shakeel Ahmed*[a], Mohammad A. Ali[a] and Nadhir Al-Baghli[b]

[a]*Center for Refining and Petrochemicals, King Fahd University of Petroleum & Minerals, Dhahran 31261, Saudi Arabia *Email:* shakeel@kfupm.edu.sa

[b]Department of Chemical Engineering, King Fahd University of Petroleum & Minerals, Dhahran 31261, Saudi Arabia

Abstract

ZSM-5 zeolites having different Si/Al ratios and modified with phosphorus were synthesized and coated over ceramic monolith structures using a dip-coating process. The prepared catalysts were characterized by SEM-EDX and BET methods to determine the morphology and surface characteristics. The nano-ZSM-5 powder and nano-ZSM-5 coated on the monolith were evaluated for methanol to propylene reaction in a fixed-bed reaction system with online GC. The nano-ZSM-5 having three different silica to alumina ratios of 30 80 and 280 were screened for methanol to propylene (MTP) reaction. It was found that nano-ZSM-5 having silica to alumina molar ratio of 280 has highest propylene selectivity. This shows that the propylene selectivity increased with increasing silica to alumina ratio of the zeolite. The zeolite (280) was then coated over the monolith and tested for MTP reaction. It was found that higher propylene selectivity was achieved with nano-zeolite coated on monolith as compared to pelletized nano-ZSM-5. This is because of the fact that monolith-based reactor exhibited very low diffusional limitations in the catalyst particles and much higher catalytic efficiency. For 100% methanol conversion, the amount of zeolite used in monolithic catalyst was several times less than the powdered catalyst.

Paper No. 18

Hierarchical nano-consortium catalyst of (Ni, Fe, Co)-Bi-O/support for oxidative dehydrogenation of butane to butadiene

G. Tanimu [1], B. R. Jermy [2], B.A. Abussaud [1], S. Asaoka [2], S. Al-Khattaf [1, 2]

1. *Department of Chemical Engineering, King Fahd University of Petroleum & Minerals, Dhahran 31261, Saudi Arabia*
2. *Center of Research Excellence in Petroleum Refining and Petrochemicals, King Fahd University of Petroleum & Minerals, Dhahran 31261, Saudi Arabia*

Abstract

Oxidative dehydrogenation of n-butane to butadiene over hierarchical nano-consortium catalyst: (main metal: Ni, Fe, Co)-Bi-O/support depends on the main metal species/composition and the support species. As for main metal effect, the catalytic performances over (main metal: Ni, Fe, Co)-Bi-O/Al_2O_3 catalyst depend on the main metal and the composition. In the case of the binary Ni-Fe, Ni-Co and Fe-Co systems, Fe and Co worked to improve Ni for the butadiene selectivity and n-butane conversion, respectively. The ternary Ni-Fe-Co system showed the highest butadiene selectivity at the highest n-butane conversion with good catalytic stability as the result due to double improvements by Fe and Co. As for support effect, the catalytic performances over Ni-Bi-O/SiO_2 catalyst strongly depend on the SiO_2 species. The order of the butadiene selectivity as support is Formed SiO_2 = SiO_2 sol > SiO_2 gel > MCM-41 > MFI-silicate, where Formed SiO_2 catalyst show the highest 2nd step dehydrogenation selectivity. The Bi/Ni reducing to 0.14 in Ni-Bi-O/SiO_2 catalyst effects positively on the butadiene selectivity, while Bi in Al_2O_3 catalyst is required as Bi/Ni=0.42. The acid and base sites preferably adjusted by hierarchical nano-particle cohabitation of main metal (Ni, Fe, Co) oxide, Bi_2O_3 and support cooperate to accelerate butadiene selectivity at 1st and 2nd step dehydrogenations.

Proceedings 26th Saudi Japan Annual Symposium
Technology in Petroleum Refining & Petrochemicals
KFUPM Dhahran, Saudi Arabia Nov. 7-8, 2016

Paper No. 19

Sales gas storage in Saudi Arabia

Rashid Al- Othman
Saudi Aramco R&DC

Abstract

This paper discusses the efforts that Saudi Aramco's Technology Oversight and Coordination has made during the past five years to demonstrate an adsorption-based process or facility that withdraws sales gas from the master gas system pipeline during off-peak demand night hours of summer days and deliver it to power generation plants during peak demand hours of summer days, to meet energy load variation in the Kingdom of Saudi Arabia. This paper will focus on the main activities that have led to completing of a feasibility study to demonstrate this process: 1- competency development for estimating performance of commercial activated carbon as storage material for sales gas, 2-development of a novel method (technology) for a facility that stores and deliver up to few billion cubic feet of sales gas, 3-completion of a pilot test for this technology, and 4- development of new production route for an adsorbent with superior storage properties. As background data, the history of sales gas production in the Kingdom as well as the different options for sales gas storage will also be included in the discussion.

Paper No. 20

Modeling HIC growth in carbon steels: Current and future work at Saudi Aramco R&DC

K.U.R.M. Shibly, A. Traidia
Research & Development Center, P.O. Box 62, Saudi Aramco, Dhahran, Saudi Arabia

Abstract

A significant number of oil and gas assets operate in wet sour environments, with increasing levels of H_2S content. Several vital components of these assets were originally made of non HIC-resistant steels, so called 'dirty steels.' These steels, of older manufacture, contain many MnS inclusions which make them sensitive to different kinds of Hydrogen Embrittlement damage, since the MnS inclusions act as nucleation sites for Hydrogen Induced Cracks (HIC) which may then coalesce to form Step-Wise Cracks (SWC).

The current integrity management strategy is based on frequent NDT inspection of all HIC and SWC affected equipment. The results of the NDT inspections are then analyzed with respect to the fitness for service criteria specified in API-579 standard to decide on whether (1) continued safe operation, (2) pressure de-rating and continuous inspection or (3) immediate shutdown and repair shall be considered. This approach, which relies primarily on systematic frequent NDT inspections can be too conservative or result in high inspection costs.

In view of supporting decision –making regarding the frequency of NDT inspection and developing a proactive rehabilitation/replacement program of HIC-damaged assets, a numerical prediction tool, named ProHIC (previously called PREDHICT), was developed [1]. The numerical tool, which is based on a mechanistic model previously published by the A. Traidia [2], makes is possible to numerically simulate the growth rate of pre-existing linear HIC (i.e. blisters) in sour service pressure equipment. This presentation will discuss the main capabilities of the software and the future work planned with regards to further developing it and its crack growth capabilities.

[1] Traidia, A., et al., Hydrogen-Induced Cracking Prediction Software. *Journal of Pipeline Engineering* (to be published in 2017).

[2] Traidia, A., et al., An effective finite element model for the prediction of hydrogen induced cracking in steel pipelines. *International Journal of Hydrogen Energy*, 2012. **37**(21): 16214-16230.

Paper No. 21

Acoustic emission monitoring of HIC growth in carbon steels: laboratory testing

H. Attar, A. Traidia, A. Enezi
Research & Development Center, P.O. Box 62, Saudi Aramco, Dhahran, Saudi Arabia

Abstract

Old manufactured steel might suffer from Hydrogen-Induced Cracking (HIC) when subjected to sour services. Acoustic Emission technique (AET) showed a potential to be utilized in order to detect the initiation and development of these cracks and others. In this test, it has been confirmed that AET has the capability of detecting these through following a methodology to distinguish the signals of HIC phenomena from other signals that are attributed to other activities in additional to the noise from the environment. In this work, three different grades of steel (X42, A106 and X65) have been tested in a sour environment to monitor that HIC development in each grade. The signals generated from these experiments have been statistically analyzed to and clustered to reflect the different physical mechanisms taking place.

Paper No. 22

Development of corrosion inspection technique using ultrasonic testing for the measurement of wall loss of inaccessible components

Satoru Shiroshita
Non-Destructive Inspection Co., Ltd, Japan

Abstract

This paper describes a newly developed ultrasonic inspection technique using two signals which are obtained by both multiple traverse and surface propagation. This technique can be applied to detect corroded area and evaluate quantitatively wall loss of inaccessible component such as pipes with support structure.

The developed technique has some advantages as follows. (1) While in conventional techniques, ultrasound propagate in either circumferential or axial direction of the pipe, this technique allows ultrasound to propagate in a diagonal direction additionally. Therefore, un-measureable area does not exist on the pipe even under the reinforcement plate around the nozzle. (2) It is possible to identify the location of corrosion, inner or outer surfaces, since two signals above mentioned are observed in this technique. (3) In conventional techniques, it is not possible to get a sufficient detectability due to sound attenuation at the corroded area, and the maximum distance between two probes is also limited. On the other hand the developed technique can obtain the stronger signal from corroded area than conventional techniques, and two probes can be placed at the longer distance.

This technique was developed as an erosion detection technique under the reinforcement plate of the power plant. Now, it can be mainly applied to a corrosion test of the pipe rack contact portion in oil refinery and petrochemical plants.

Paper No. 23

Predictive modeling of the internal localized corrosion rate in wet hydrocarbon pipelines

T. Villette, A. Sherik
Research & Development Center, P.O. Box 62, Saudi Aramco, Dhahran, Saudi Arabia

Abstract

In the oil and gas industry, one significant factor in the degradation of pipelines is internal corrosion. Pipeline corrosion generally manifests itself in the form of either uniform corrosion or localized corrosion. Uniform corrosion is easier to monitor, control and predict and therefore is not considered a major problem. On the opposite, localized corrosion is the most dangerous form of corrosion and constitutes the most corrosion failures experienced by Saudi Aramco and generally speaking by the oil and gas industry. It is therefore the main limiting factor affecting pipeline integrity. Because of its highly localized and statistical nature, pitting corrosion is difficult to simulate in the laboratory or monitor and predict in the field. These difficulties are additionally compounded by the fact that there are limited laboratory studies on pitting corrosion in sour environments mainly because of difficulties in simulating the field conditions and because laboratories have limited capabilities to run sour corrosion testing for health and safety reasons. All this makes modeling and prediction of localized corrosion rate in sour environment an important endeavor with potential high value impacts not only to the oil and gas industry but to the industrial sector at large.

This presentation briefly discusses a new numerical tool that is currently being developed by the Modeling & Prediction Team of the Oil and Gas Network Integrity Division-R&DC for the prediction of internal localized corrosion in wet hydrocarbon lines.

26th Annual Saudi-Japan Symposium – 2016

السعودية أرامكو
Saudi Aramco

Technology in Petroleum Refining & Petrochemicals

***Venue:** KFUPM Research Institute Bldg. 15; 4th Floor, Dhahran, Saudi Arabia, November 7-8, 2016*

Day One: Monday, November 7, 2016

OPENING REMARKS SESSION — ***Chairman: Dr. Sulaiman Al-Khattaf***

7:45 *Registration & Coffee Break*

8:00 *Opening Remarks*

- *H.E. Dr. Khaled S. Al-Sultan, Rector of KFUPM*
- *Mr. Eiji Hiraoka, Senior Executive Director, JCCP*
- *Dr. Keiichi Tomishige, Head Japan Delegation, JPI*

SESSION ONE MODELING/CATALYTIC REACTIONS — ***Chairman: Dr. Atsushi Ishihara***

8:30 1. **Software tools for molecular-level kinetic modeling of refinery and petrochemical reactors,** *Dr. Michael Klein, Saudi Aramco Chair Professor, KFUPM, Univ. Delaware, USA*

9:00 2. **Hydrogen production by steam reforming reactions over uniform Ni-Fe alloy nanoparticles,** *Dr. Keiichi Tomishige, Tohoku University, Japan*

9:20 3. **Novel approach to overcome mass-transfer limitation in multiphase reactions with Taylor-Couette flow,** *Mr. Mohammad Aljishi, Saudi Aramco R&DC*

9:40 4. **Supported Keggin-type heteropolyacid catalysts for isobutene oligomerization and methacrolein oxidation,** *Dr. Yuichi Kamiya, Hokkaido University, Japan*

10:00 *Coffee Break*

SESSION TWO HYDROPROCESSING — ***Chairman: Dr. Mohammed Ba-Shammakh***

10:30 5. **Hydroprocessing catalysts for heavy oil upgrading,** *Dr. Lianhui Ding, Saudi Aramco R&DC*

10:50 6. **SPH (Slurry Phase Hydrocracking) Process as a cutting-edge upgrader for upcoming diesel demand in not-too-distant future,** *Mr. Daisuke Oka, Chiyoda Corporation, Japan*

11:10 7. **CAN-15 catalyst- data review,** *Mr. Vinod Ramaseshan, Saudi Aramco P&CSD*

11:30 8. **Oil refining in a CO_2 constrained world with implications for gasoline and diesel fuels,** *Mr. Amir Abdul Manan, Saudi Aramco R&DC*

11:50 *Prayer & Lunch Break*

SESSION THREE AROMATICS/PETCOKE — ***Chairman: Dr. Rashid Al-Othman***

13:00 9. **Development of the efficient process for producing BTX from fuel oil** *Mr. Masahide Kobayashi, JX Nippon Oil & Energy Corporation, Japan*

13:20 10. **Selective dehydrocyclization reaction of n-paraffins using zeolite-oxide composite supported NiMo-based catalysts,** *Dr. Atsushi Ishihara, Mie University, Japan*

13:40 11. **Anode-grade coke from Arabian crudes,** *Dr. Bandar Solami, Saudi Aramco, R&DC*

14:00 *Day One Ends*

Each presentation includes 5-minutes Q&A

26th Annual Saudi-Japan Symposium – 2016

ارامكو السعودية
Saudi Aramco

Technology in Petroleum Refining & Petrochemicals

Venue: KFUPM Research Institute Bldg. 15; 4th floor, Dhahran, Saudi Arabia, November 7-8, 2016

Day Two: Tuesday, November 8, 2016

SESSION FOUR CATALYTIC OLEFINS — *Chairman: Dr. Yuichi Kamiya*

8:15 — *Registration & Coffee Break*

8:30 12. **Active and stable catalyst for 1-butene cracking to propylene**
Dr. Arudra Palani, KFUPM-CRP

8:50 13. **Development of highly selective zeolite based catalysts for naphtha catalytic cracking to ethylene and propylene,** *Dr. Nabil Al-Yassir, SABIC Technology & Innovation, Riyadh*

9:10 14. **A light olefin selective FCC catalyst based on Ti, Zr-inserted USY zeolite**
Dr. Yaming Jin, Saudi Aramco R&DC

9:30 15. **An integrated study on the production of light olefins from catalytic cracking of Arab crude oil fractions in different reactor systems,** *Dr. Qi Xu, Saudi Aramco R&DC*

9:50 — *Coffee Break*

SESSION FIVE CRACKING/DEHYDROGENATION/GAS — *Chairman: Dr. Abdullah Al-Malki*

10:20 16. **Study the nano-structured zeolite for hydrocarbon cracking**
Dr. Emad Shafi, Saudi Aramco R&DC

10:40 17. **Zeolite supported structured catalysts for methanol to propylene conversion**
Dr. Shakeel Ahmed, KFUPM-CRP

11:00 18. **Hierarchical nano-consortium catalyst of (Ni, Fe, Co)-Bi-O/support for oxidative dehydrogenation of butane to butadiene,** *Mr. Gazali Tanimu, KFUPM-CRP*

11:20 19. **Sales gas storage in Saudi Arabia,** *Dr. Rashid Al-Othman, Saudi Aramco R&DC*

11:40 — *Prayer & Lunch Break*

SESSION SIX CARBON STEELS/CORROSION — *Chairman: Dr. Abdulaziz Al-Saadi*

13:00 20. **Modeling HIC growth in carbon steels: current and future work at Saudi Aramco R&DC**
Mr. Kaamil-Ur-Rahman Shibly, Saudi Aramco R&DC

13:20 21. **Acoustic emission monitoring of HIC growth in carbon steels: laboratory testing**
Mr. Hussam Attar, Saudi Aramco R&DC

13:40 22. **Development of corrosion inspection technique using ultrasonic testing for the measurement of wall loss of inaccessible components,** *Mr. Satoru Shiroshita, Non-destructive Inspection Co., Japan*

14:00 23. **Prediction of the internal localized corrosion rate in scrapable transmission pipelines**
Mr. Thibault Villette, Saudi Aramco R&DC

14:20 — ***Closing Remarks, Symposium Ends***

Each presentation includes 5-minutes Q&A — ***Program: October 26, 2016***

King Fahd University of Petroleum & Minerals
Dhahran 31261
Kingdom of Saudi Arabia

ISBN 978-1-5108-3306-7

9 781510 833067

CURRAN ASSOCIATES INC.
proceedings.com

New Zealand Society of Animal Production Conference (NZSAP 2023)

The land, the water, the people.

Proceedings of the New Zealand Society of Animal Production
Volume 83

Rotorua, New Zealand
14 - 16 November 2023